D1733034

Unbroken Spirit: The Wild Horse In The American Landscape is published on the occasion of a major exhibition of the same title. Organized by Charles R. Preston, curator of Natural History at the Buffalo Bill Historical Center, the exhibition was shown at the Historical Center, July 23—October 31, 1999; and at the Western Museum of Wildlife Art, November 19, 1999—April 9, 2000.

Library of Congress Catalog Card Number:
99-072857
International Standard Book Numbers:
0-931618-62-2

Unbroken Spirit

THE WILD HORSE IN THE AMERICAN LANDSCAPE

Edited by Frances B. Clymer and Charles R. Preston

Charles R. Preston

Lynne Bama

Daniel L. Flores

Linda Coates-Markle

B. Byron Price

Photography by Gary Leppart, Dewey Vanderhoff,
John Eastcott and Yva Momatiuk

Cover: John Eastcott and Yva Momatiuk
Project Coordinator: Renee Tafoya
Designer: Jan Woods Krier

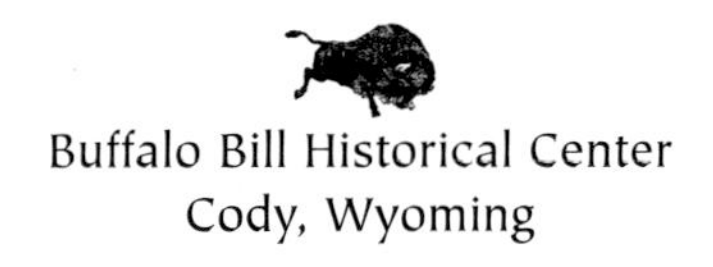

Buffalo Bill Historical Center
Cody, Wyoming

Contents

Left: Gary Leppart. Photography credits from top (details): Gary Leppart, Dewey Vanderhoff, Gary Leppart, John Eastcott and Yva Momatiuk, Buffalo Bill Historical Center.

Gary Leppart

Introduction

Charles R. Preston

Curator of Natural History
Buffalo Bill Historical Center

The road, if indeed it could be called a road, was absolutely brutal. We had been in the saddle of the old Suburban for more than eight hours, and this last stretch of dry, gully-ridden, boulder-strewn landscape was beginning to take a serious toll on certain parts of my anatomy. The scenery, however, was spectacular. Stark, deep-red rocks and soil contrasted dramatically with the rich ever-green hues of piñon pine and juniper. Blooming prickly pear, claret cup, and other cactus species lent splashes of bright color, and widely spaced patches of rabbitbrush, needle-and-thread, and various species of bluegrass completed the picture. This was rough country, typical of much of the landscape in drier regions of the Rocky Mountain West. The year was 1991, and I was about to embark on an adventure that would dramatically influence my perspective on interpreting natural history in a museum setting.

I was accompanied on this fateful trip by a friend and professional colleague who was intimately familiar with this sparsely populated region of northwestern Colorado. I had recently begun a stint as chairman of the Department of Zoology at the Denver Museum of Natural History, and was anxious to explore the nether regions of Colorado. We were on a section of land managed by the U.S. Bureau of Land Management in the Piceance Basin, in Rio Blanco County. Among other attractions, we were hoping to catch a glimpse of some of the wild horses known to inhabit the region. I was not prepared for our sudden, startling success! As we rounded a sharp curve traversing the pygmy woodland, the landscape lay uncluttered before us in a broad, rocky depression. On the edge of a bluff bordering the depression stood a large chocolate-and-white paint horse, with its long, white mane dancing in the breeze. The scene was almost surreal, and I had to blink a couple of times to assure myself that this was no mirage. As we watched, two other horses came into view briefly, before all three turned away from the bluff's edge, and disappeared from view.

Now, I should admit up front that I never gave much thought to wild horses before that day. I knew that several centuries after ancestors of the modern wild horse became extinct in North America, domestic horses were introduced to this continent by early Spanish explorers. These animals represented a stock that had undergone intensive selective breeding in Europe and Asia for thousands of years. Many of the horses escaped human control, and survived and reproduced in the American landscape, forming closely-knit bands. Although the term "feral" is generally applied to domestic livestock that becomes free-roaming, most Americans simply refer to these animals as "wild". Some wild horses roaming the West today are direct descendants of the Spanish stock introduced to this continent more than 400 years ago. They have been joined by escaped or abandoned horses of various breeds through the years.

As a wildlife biologist, I tacitly placed modern North American wild horses in the same category as European starlings and feral hogs (razorbacks); alien animals that roam free in 20th century North America only because of human introduction and neglect. My chief concern was for the fate of native wildlife that shared the range with wild horse herds. But I must confess that the sight of those horses in the Piceance Basin struck an emotional chord. There was something that seemed fitting and, well . . . natural, about the horses running unfettered in the western landscape. Maybe it was the long-repressed image of *Fury* or other television programs and movies I was addicted to in my youth that linked the wild horse with the spirit of the West. At any rate, I was suddenly struck with ambivalence about the wild horse. On one hand, I was concerned about the potential impact of an introduced, feral animal into native ecosystems, and on the other, I had to admit that the presence of the wild horse added a dramatic, even romantic, dimension to the landscape. I soon found out that my own ambivalence reflected a much broader and more intense polarization among western citizens.

Later during that same field expedition, my colleague and I visited the small town of Meeker, Colorado, about 40 miles northeast of the site where we encountered the wild horses. The local watering holes in Meeker attract a diverse group of patrons, including local ranchers and farmers, together with hikers, mountain bikers, wildlife watchers, and other outdoor enthusiasts from throughout the region. When we stopped into one of the local saloons, we got an earful about wild horses. Several of the local ranchers expressed their opinion that wild horses offered serious competition to cattle and other domestic livestock and should be removed from the open range. At the same time, many of the local and visiting recreationists argued passionately that wild horses had a positive ecological impact on the land, had come home to take their rightful place in the Wild West, and should be protected and nurtured at all costs. As the debate heated up, I noticed that few listeners took a middle stance between the two polar positions. Most people, offended by arguments advanced by one side or the other, began allying themselves with the opposite camp. The arguments became less about facts of ecology and economy, and more about human emotions and alliances. Many folks appeared to choose a particular side, not because of any strongly held convictions about wild horses, but because they wanted to distance themselves from people they didn't want to identify with on the other side. The experience reinforced in me an axiom I had related many times to my university students in wildlife management years before: Modern natural resources management is as much about understanding people as it is about understanding natural resources. This reminder jolted me into recognizing a glaring oversight in most natural history museum-based programming about contemporary topics; humans were usually left out of the story!

I returned to the museum with the idea of developing an exhibition about wild horses that would incorporate human cultural history, values and perceptions equally with horse evolution, ecology, and behavior. But bureaucratic obstacles to a multidisciplinary exhibition can be formidable in a large, traditional natural history venue, and there was more than enough to keep me busy without taking on an uphill campaign to squeeze a new exhibition into our schedule. The subject stayed buried deep in my file drawer until June of 1998, when I assumed my position at the Buffalo Bill Historical Center. It was then that Executive Director B. Byron Price called me into his office and informed me that one of my first assignments as natural history curator would be to lead the development of an exhibition on wild horses that was scheduled to open in the summer of 1999!

In early discussions with Associate Director and exhibit designer Wally Reber, about the trials and tribulations of wild horses in North America and the renegade image they portray to many people, a title for the exhibition emerged: *Unbroken Spirit: The Wild Horse in the American Landscape.*

I can't think of a subject that better epitomizes the complex, dynamic relationship between humans and their environment in the American West. The presence of wild horses in the American landscape has produced a wide range of responses among those with whom they share the land. Some people perceive these horses as threats to their livelihoods or to the fragile ecosystems of the West and wish to see their numbers reduced or to have them eliminated entirely. Others work tirelessly to maintain herd populations and shelter mustangs from any harm. Still others, who live removed from the lands mustangs occupy, see them simply as magnificent symbols of the wildness and freedom of the mythic West.

Our goal with *Unbroken Spirit*, the exhibition, is not only to present the biological and historical facts about America's wild horses, but to convey a sense of the human emotions invested on all sides of the wild horse story. We've strived to engage, evoke, and provoke, as well as to inform. The texture of the exhibit, as rich and multi-dimensional as the subject itself, incorporates inspirational artwork, breathtaking contemporary photography, historical photographs and documents, cultural artifacts, and dramatic taxidermy.

One of the most exciting and important opportunities presented by *Unbroken Spirit* is the development of ancillary products that will live well beyond the exhibition itself. Among these is an interactive CD-ROM, videotaped proceedings of a symposium held in conjunction with the exhibition, and this catalog.

This volume presents the scholarship, ideas, and passions of five individuals from vastly different backgrounds. Any multidisciplinary project is inherently ambitious, and to blend science and humanities is particularly challenging. But if we are to better understand the relationship between humans and nature, and find satisfactory solutions to the growing number of environmental problems confronting us, we must explore both how nature works and how human perceptions and desires develop. The greater our perspective, the greater our ability to understand.

In these essays you will find broad areas of overlap, but you will also discover the most important truth about wild horses: their place and significance in the American landscape is subjective, dependent on the "eyes" of the beholder. The story that emerges is complex, and serves as much as a mirror on our evolving American culture as it does a microscope on the animal at center stage. ■

1 An Integral Part of the Natural System

Charles R. Preston

It is not merely soil, nor plant, nor animal, nor weather which we need to know better, but chiefly man himself.

—Paul Sears

The pungent aroma of sagebrush fills the air as you and your horse crest another of the seemingly endless hills in the rugged, western landscape. No trees interrupt the scene stretched out before you; a deeply textured tapestry of tan and red earth punctuated by dappled gray and brown boulders and pale green splashes of sagebrush. You hear no telephones ringing, no computer keyboards clicking, no automobile horns honking and no internal combustion engines rumbling. Indeed, the only sounds you hear are the deep, rhythmic breaths of your mount and the lonesome whistle of the pervasive Wyoming wind. Riding that wind 100 feet above you is the only sign of life immediately apparent in this vast, shrub-steppe wilderness—a golden eagle with a wingspan longer than basketball great Michael Jordan is tall. But when you scan the surrounding landscape with binoculars, a herd of pronghorn becomes visible, and a lone coyote, nose to the ground, emerges from a distant arroyo. Then suddenly you become aware of another movement 300 yards to your right! Your horse stands at attention, ears cocked in the direction of the movement. First one, then two forms take shape. After a few seconds, more than 15 animals of varying colors, shapes and sizes are in your field of view. They are aware—and wary—of you, but they do not flee immediately. After a few seconds, one of them turns away with a loud snort and leads

Left: Gary Leppart

the others out of sight at a gallop. These animals seem familiar to you, yet vaguely foreign in their movements and behavior. They are horses, the same species (*Equus caballus*) as your domesticated mount.

But the horses you have encountered in this Wyoming rangeland are free-roaming horses that have never been domesticated. Some of them may be descended from horses that roamed this range more than 400 years ago after escaping domestication. Though scientists generally use the term feral to refer to free-roaming descendants of domestic animals, the public has embraced these animals as wild. Indeed, the free-roaming horse has come to symbolize the spirit of wildness itself to many Americans—so much so that the federal government has granted explicit protection to the wild horse as a living icon of the frontier West and an important component of the environment.

Not everyone agrees, however, that the wild horse deserves its revered and protected status. After all, this symbol of wild America is descended from immigrants introduced to this continent as domestic animals. The recent ancestors of contemporary wild horses were renegades from domestication. Nonetheless, few animal species, alien or native, have captured the imagination and passion of so many Americans. To understand how this creature has become so entrenched in our psyche and landscape, it is helpful to briefly explore the lineage and history of horses in North America and examine the ecological niche now occupied by the wild horse in the American West. The expedition reveals perhaps as much about our own species as it does about the horse.

It is apparent from the known fossil record that North America is the seat of early horse evolution. About 60 million years ago, the diminutive (about 12 inches tall) and much-celebrated Dawn Horse, *Hyracotherium* roamed the continent. Through millions of years and untold environmental changes, ancestral horses took on a bewildering number of forms. Roughly three million years ago, the genus *Equus* emerged in the American landscape. Early members of this genus were similar, at least in terms of skeletal characteristics, to the modern horses. Ample evidence suggests that early *Equus* species were widespread and abundant in North America and extended their range into Eurasia via the Bering land bridge. The range extension of *Equus* beyond North America proved fortunate for descendants, because horses became extinct on this continent some 8,000–12,000 years ago. They disappeared as part of a massive extinction of North American megafauna (e.g., mammoth, rhinoceros, sabertooth cat, camel, short-faced bear, etc.) at the end of the last great Ice Age. Climate change, disease and human hunting pressure may all have played a part in the extinction, though the relative importance of proposed causes remains the subject of heated scientific debate.

Early Spanish explorers introduced the modern horse to North America as domestic livestock in the 16th century. Though it might be argued that this marked the triumphant return of the horse

Left: John Eastcott and Yva Momatiuk

W. H. D. Koerner (1878-1938), *Horses, 1927-1933, Breaking Colts, Phil Spear Ranch, Lodge Grass, MT*, black and white photograph. Buffalo Bill Historical Center, Cody, WY. Gift of the artist's heirs, W. H. D. Koerner III, and Ruth Koerner Oliver.

to its former homeland, the horses introduced here by Europeans had undergone intensive selective breeding for thousands of years, and represented a stock quite different from any North American ancestor. As a domestic animal, this import profoundly influenced American cultures in the 18th and 19th centuries.

Dewey Vanderhoff

Inevitably, some horses from this alien stock escaped domestication or were abandoned and they thrived in the arid, open rangeland. Descendants of the old Spanish stock were joined in many areas by escaped/abandoned horses brought into the frontier by trappers, settlers, miners and other immigrants, and by the early 1800s, wild horses occupying western North America may have numbered in the millions. But the open range began to shrink, as cattle, sheep, fences and farms spilled out into the Great Plains. Bands of wild horses continued to survive in remote desert and semi-desert regions of the West, but numbers began to decline.

This trend continued into the 20th century, as wild horses were shot to reduce competition with domestic livestock and rounded up by the tens of thousands for use as farm and ranch horses and mounts for cavalry in foreign wars. Wild horse populations actually grew during the Great Depression, as large numbers of domestic horses were abandoned by owners who could no longer care for them.

With the emergence of European markets for horse meat, and domestic and foreign markets

John Eastcott and Yva Momatiuk

for pet food and chicken feed, wild horses again began disappearing in great numbers from the western range. Mustanging, the business of capturing and transporting wild horses for profit, became a thriving enterprise for some. Their methods were often brutal. When knowledge of this brutality reached the general public, a wave of outrage swept the nation and a movement to protect wild horses was born. Leading the charge was Mrs. Velma Johnston, who became known as Wild Horse Annie. Largely due to her efforts, Congress passed legislation in 1959 to prohibit the use of motorized vehicles for capturing or harassing wild horses. Ultimately, the public outcry championed by Wild Horse Annie, and fueled by the passion of a growing number of wild horse advocacy groups, led to the passage of the Wild Free-Roaming Horse and Burro Act in 1971.

The wild horse had clearly touched a nerve in the American psyche. The Act declared wild horses and burros as ". . . living symbols of the historic and pioneer spirit of the West," and that they ". . . shall be protected from capture, branding, harassment, or death." Furthermore,

Congress declared that wild horses and burros ". . . are to be considered in the area where presently found, as an integral part of the natural system of the public lands." Responsibility for protecting wild horses and burros fell to the Bureau of Land Management and, to a lesser degree, the National Park Service and U. S. Forest Service, as stewards of public lands occupied by these animals.

It did not take long for wild horse populations on public lands to increase dramatically, and

John Eastcott and Yva Momatiuk

Charles A. Belden, *Cowboys Trailing Cattle*. Buffalo Bill Historical Center, Cody, WY. Charles Belden Collection.

concerns arose regarding the effects of wild horses on deer, elk, bighorn sheep, and domestic livestock sharing the same range. "Natural" predation is not a significant source of mortality for most wild horse populations. Stockgrowers and other residents, with the support of federal and local governments, have successfully reduced or eliminated large predators from much of the western rangeland, and sabertoothed cats and dire wolves have long since vanished. With the absence of natural predation, some agricultural users of public lands called for massive reductions of wild horse herds by virtually any means necessary. Diametrically opposed were vocal wild horse advocates, who felt that the estimated numbers of wild horses were inflated and that no herd reduction was justified. Capturing the middle ground were those who supported the continued presence and protection of the wild horse on public lands, but recognized a need for judicious and humane herd reductions to ease impact on an ecologically stressed landscape. These three viewpoints reflect marked differences in personal interests, values, and perceptions.

Eventually, Congress authorized the Adopt-A-Horse program, administered by the Bureau of Land Management. Under this program, horses from free-roaming, overpopulated herds were removed from public lands, and provided for adoption by private, non-commercial owners. The program continues today, though not without problems and controversy. Unfortunately, there is no universally accepted protocol for judging whether a wild horse herd is overpopulated, or by how much it should be reduced. Even if we could agree on a standard on which to judge over-population, it may not be possible to remove and adopt enough horses to effectively manage populations.

Sound management depends on sound information about wild horse ecology. Unfortunately, the relationship between wild horses and potential range competitors varies with geographic area, season, weather and many other factors. Wild horses are largely descended from domestic strains selected thousands of years ago for their abilities to thrive in arid, semi-desert conditions, similar to those found on current wild horse ranges. Studies specifically designed to quantify ecological relationships among wild horses, native wildlife and domestic livestock on these ranges are sparse. Nonetheless, some useful information has emerged from widely scattered studies. Horses, like cattle, are primarily grazers, rather than browsers. Depending on site characteristics and season, however, their dietary overlap can vary a great deal. The potential for overlap is generally greatest in winter, when grasses dominate the diets of both species. Horses tend to eat proportionately more than cattle and are able to nip grasses closer to the ground. Nonetheless, domestic livestock typically outnumber wild horses in most areas where they coexist, and consume the lion's share of the forage available. There is generally less dietary overlap between wild horses and native wildlife than between wild horses and cattle, but the potential for competition exists. Pronghorn and mule deer are common residents in wild horse ranges, elk occupy many areas seasonally, and bighorn sheep share some sites, such as the Pryor Mountain Wild Horse Range in northern Wyoming. Bighorns and horses share a very similar diet during spring and summer months. Elk, deer and pronghorn browse on shrub and tree species much of the year, but also share a similar diet with wild horses during spring and summer. In "ecological crunch" times, when food or water are scarce, wild horses could negatively impact domestic livestock and native wildlife. The magnitude of the impact would depend largely on the size of the horse population.

Thus, the modern wild horse may be characterized as an introduced animal that has escaped human bondage and potentially impacts native wildlife and domestic livestock on the public rangelands of the American West. Among its potential competitors, the wild horse stands alone as the species with virtually no monetary value assigned to it. Native deer, pronghorn, elk and bighorn sheep are all game animals, and hunters pump significant currency into western communities. Cattle and sheep may be non-native animals that compete with native wildlife, but they are highly valued as domestic livestock and a source of personal income for stockgrowers.

John Eastcott and Yva Momatiuk

Livestock ranching continues to play a dominant role in the economy of the West. Cattle on public lands outnumber wild horses by more than four to one, and stockgrowers contribute a strong voice to the dialogue concerning public land use.

Many introduced wildlife species with no clear economic value, such as the Norway rat, English sparrow and European starling, are generally held in low esteem by the American public and are frequently persecuted as nuisance species. Yet in spite of its introduced status, its lack of any significant economic value, and its role as a potential competitor with native wildlife and domestic livestock, the wild horse remains a widely revered and protected inhabitant of the American landscape.

Perhaps the wild horse has come to represent a vision of how many Americans would like to see themselves: a rakish renegade, choosing the delights and challenges of freedom to the reins of control. The modern horse traces its ancestry back to another land, and like many Americans, the wild horse has broken bonds of repression to find freedom in this land. We have chosen to view the wild horse differently than the Norway rat or European starling. Our view may be influenced to some degree by the fact that ancestors of the modern horse once roamed this continent before becoming extinct. Most certainly our view is influenced by our perception of the wild horse as a charismatic animal. The designation of the wild horse as "an integral part of the natural system of the public lands" is certainly more about human cultural perceptions and values than biology. The last natural system in North America that included horses as integral also included mammoths, camels, sabertoothed cats, and more.

The wild horse fits nicely into the vision of the western landscape embraced by many Americans. The future of the wild horse, indeed the future of the American landscape and all of its inhabitants, is inextricably linked to human cultural values. Wildness, after all, has become a relative term in modern America. It exists under the auspices and, to some extent, by the design of one species, *Homo sapiens*. The wild horse, a celebrated symbol of unbridled freedom, may thus be viewed as merely one of many free-roaming species captive to the ever-evolving human vision of how we want our world to look. If the wild horse survives in the American landscape, it will be because, as we strive to control more of our world and tame its frontiers, we assign increasing value to the unbroken spirit that struggles against our control and reminds us of the real challenges and fading glory of wildness.

As western wildlands continue to decline, however, and if the welfare of wild horses comes into serious conflict with the welfare of native wild species, we may be forced to choose which vision of wildness we prefer and which species we value more as an integral part of the natural system. ■

Gary Leppart

SOURCES AND SELECTED READINGS

Charles R. Preston

Anthony, D. W. 1997. "Bridling Horse Power: the Domestication of the Horse." In *Horses Through Time*, edited by S.L. Olsen. Boulder, CO: Roberts Rinehart Publishers for Carnegie Museum of Natural History.

Berger, J. 1986. *Wild Horses of the Great Basin: Social Competition and Population Size*. Chicago, IL : University of Chicago Press.

Coates, K. P. and S. D. Schemnitz. 1994. Habitat use and behavior of male mountain sheep in foraging associations with wild horses. *Great Basin Naturalist* 54:86-90.

Crane, K. K., M. A. Smith and D. Reynolds. 1997. Habitat selection patterns of feral horses in south-central Wyoming. *Journal of Range Management* 50:374-380.

Hubbard, R. E. and R. M. Hansen. 1976. Diets of wild horses, cattle and mule deer in the Piceance basin, Colorado. *Journal of Range Management* 29:389-392.

Hulbert, R. C. 1997. "The Ancestry of the Horse." In *Horses Through Time*, edited by S.L. Olsen. Boulder, CO: Roberts Rinehart Publishers for Carnegie Museum of Natural History.

Kirkpatrick, J. F. 1994. *Into the Wind: Wild Horses of North America*. Photography by Michael H. Francis. Minocqua, WI: NorthWord Press.

Krysl, L. J., M. E. Hubbert, B. F. Sowell, et al. 1984. Horses and cattle grazing in the Wyoming Red Desert I: food habits and dietary overlap. *Journal of Range Management* 37:72-76.

McFadden, Bruce J. 1992. *Fossil Horses: Systematics, Paleobiology, and Evolution of the Family Equidae.* Cambridge [England]; New York: Cambridge University Press.

Simpson, George Gaylord. 1951. *Horses: the Story of the Horse Family in the Modern World and through Sixty Million Years of History.* New York: Oxford University Press.

The Wild Free-Roaming Horse and Burro Act of December 15, 1971. PL 92-195, 85 Stat 649.

Thomas, H. Smith. 1979. *The Wild Horse Controversy.* S. Brunswick, NJ: A. S. Barnes.

About the Author

Charles R. Preston is the founding curator, Draper Museum of Natural History, Buffalo Bill Historical Center. Prior to assuming his current position in June 1998, he was the chairman of the Department of Zoology, Denver Museum of Natural History (1990-1998) and assistant/associate professor of biology, University of Arkansas, Little Rock (1982-1989). He completed his doctorate in zoology, with an emphasis in wildlife ecology, at the University of Arkansas, Fayetteville in 1982. Dr. Preston has conducted fieldwork in Belize, Ecuador, and the Galapagos Islands, as well as throughout much of North America, and has authored a book on the Red-tailed Hawk (available spring 2000) and more than 40 technical and popular articles on a wide range of topics. His current research interests focus on the impact of human activities on landscape and wildlife in the western interior of North America.

John Eastcott and Yva Momatiuk

2 Doing Time

Lynne Bama

Mounted on the gateposts of the Wyoming Honor Farm are two metal replicas of the state's emblem—a rider on a bucking bronc—transformed into folk art with painted details. The signs are appropriate, for this 800-acre facility currently houses not only 154 inmates but some 60 wild horses in the process of being domesticated.

Once I get past the checkpoint, whose sign warns visitors that they may be subject to search (I am relieved to be waved on through), I drive into a yard with clipped lawns, beds of cosmos and marigolds, and elderly shade trees. Most of the buildings are elderly, too. The farm was established early in the century, and pigs, dairy cattle, and chickens were once raised here. An impressive old barn, now used for storage, stands behind the administration building.

It is only when I notice the rows of corrugated metal buildings that the real purpose of this place becomes clear. This compound surrounded by a high chain link fence is where the inmates spend their nights.

Many of the men who live here drifted into Wyoming during one of its several energy booms, got into trouble through drugs or alcohol, and thus became long-term—if not exactly willing—residents. They've earned the privilege of coming to the farm from the state penitentiary in Rawlins by good behavior.

The horses came here from U.S. Bureau of Land Management corrals at Rock Springs, after being rounded up from the nearby Red Desert, where their only crime was their too-robust fertility. Wild horse herds reproduce at incredible rates—as high as 20 percent a year—and since they have virtually no predators but people, they soon threaten to destroy every range they inhabit. In 1998, over a thousand of them had to be captured in Wyoming alone.

I sign in at the administration building, which reminds me of a public school. The hall in front

Photo courtesy of the *Riverton Ranger*, Riverton, WY.

of the office is busy, with inmates mopping the floors and cleaning, people bustling to and fro with papers and cups of coffee. I sit on a wooden bench and wait for Mike Buchanan, the manager of the wild horse training program.

Prison training programs for these animals were begun in 1986, after a national advisory board recommended them as a way to make the West's thousands of surplus wild horses more adoptable. I can't help thinking, though, that whoever envisioned the juxtaposition of an animal that is to most Americans a powerful symbol of freedom with penal inmates had a quirky imagination. Give wild horses to convicts to shape into domestic creatures, as though they were license plates? But on second thought, maybe the idea is not so farfetched. After all, who could better teach a wild animal the arts of captivity?

With a sudden clumping of boots, a tall, cowboy-hatted figure fills the doorway to my left. Buchanan looks like the Marlboro Man with about six inches and a lot of weather added on.

He offers me a cup of coffee, and while I drink it I ask him how he got into this business.

"I've always been on a ranch," he tells me, "broke horses and cowboyed all my life." But as he got on in years he realized that "there's nobody that's going to hire an old cowboy like me." So he took a job as a guard at the state penitentiary. After he had been there eight years this position with the horses opened up at the Honor Farm, and he jumped at the chance.

"I just happened to hit a big red rose," he says.

We climb into one of farm's pickups and drive north to the small rise covered with corrals, called Horse Hill. Pieces of inner tube dangle from the metal fences here. Unbroken animals are first tied to these, so they won't injure themselves struggling with an unforgiving rope.

I've seen wild horses in pens before. They mill nervously, trying to stay as far away from people as the bars let them. Their coats are dusty, their tails drag the ground, and they peer out at the world through long, tangled forelocks. The foals try to hide behind their mothers and the stallions, edgy in their confinement, lunge and snap at their penmates.

Nothing like that is going on here.

Buchanan brings the truck to a stop near a man who is washing down a black horse. The animal steams in the cold air as he rubs its back and works the flow of water from a hose through its tail. It watches its groomer alertly, but with more interest than alarm. Other, saddled, animals appear to be asleep in the morning sun. This batch of horses has been on the farm for three and a half months.

The transformation from terrified captives to placid human companions in so short a time surprises me even more when Buchanan tells me that most of the prisoners he works with know nothing about horses when they first come here.

"We go through the first-grade lessons," he says. They learn how to shut a gate, feed a horse hay, and keep track of gloves, boots, and hats. In spite of the implication of the bronc figures at the farm's entrance, the training methods here do not include anything resembling a rodeo.

"We've learned so much about the horse in the last six to eight years," Buchanan says. "We don't "Wild West" them anymore."

One of the main objects of this program, he says, is to get the inmates to bond with the horses—by petting them, leading them around, and rubbing them down. Only when they feel comfortable with the animals do they progress to saddles, bridles, and riding. The horses and the inmates pretty much learn all this together.

Buchanan climbs out of the pickup and takes me over to a small corral where a young man in a red and white feedlot cap is riding a bay gelding. Every time the horse gets to a puddle on one side of the corral it shies. Buchanan climbs over the fence and stands in the mud, calling instructions to the rider. A minute or two later the horse calms down.

Back outside, Buchanan tells me this inmate lacks self-confidence and needs support.

Photo courtesy of the Riverton Ranger, Riverton, WY.

Photo courtesy of the *Riverton Ranger*, Riverton, WY.

Wanting to show me a different style with horses, he calls another man over to a corral where a little blue roan is saddled.

This inmate approaches the animal with confident ease, but his mount is jumpy and nervous. It flattens its ears and humps its back as it trots around the small enclosure. Buchanan calls to the rider not to keep pulling on the rein after the horse turns. The animal grows quieter as we watch.

"How's it going now?"

"Getting better," the rider replies.

"You're having trouble because you're doing things wrong," Buchanan says bluntly.

The inmate dismounts. "Yessir," he says to his saddle. I can feel a wave of his anger roll out through the bars of the fence, but Buchanan seems unconcerned.

An inmate named Rick leads a foal up to me.

"What's his name?" I ask.

"Odie. Watch out for him—he bites."

The little animal doesn't look very vicious. He nuzzles my hand and lets me scratch him behind the ears. Rick tells me, somewhat apologetically, that he's working with the foals because he got bucked off and tore a groin muscle recently.

He points to a distant corral which holds the "zeroes"—the horses that can't be trained at all.

"What happens to them?" I ask. Rick isn't sure. He guesses that the BLM turns them loose again. It sounds like wishful thinking.

Now a straggling line of men in yellow shirts, new arrivals from Rawlins, is led past Horse Hill, getting a look at the farm. Soon they will be allowed to choose whether they want to help raise cattle, grow hay and corn, work in the shop, do carpentry, or train horses. They look dazed to be out here under this huge blue sky, among the open fields. Some of them are so young and vulnerable looking, my heart is wrenched with pity.

"What did you do, to get put in here?" I want to ask. And it occurs to me that every inmate I encounter today must realize what I'm wondering. The existence of this barrier between us is something I hadn't anticipated, but now I realize that it casts a shadow over every contact between inmates and the rest of the world.

"We're the outcasts here," Buchanan tells me. "If we're going to do our job, we have to break that barrier." And the horse, he says, is the way he does it. "The horses are the key to this teaching."

Buchanan would rather not know what an inmate's crime was.

"It's hard to look a baby raper in the eye and tell him, 'You're doing a good job.' " But he adds that all he has to do is watch a man in the corral with horses to figure it out.

"The horses let us know what he's done," he says.

"If I had my choices, I'd hire all killers," he adds. He likes their aggressiveness, their energy, their need to be in control.

"When we get a tough guy who wants to go to Horse Hill," he says, "you want to be real soft with him. Let him have some power." Buchanan even allows an inmate like this to get—or appear to get—some control at first, looking the other way as he bullies others into shoveling manure and giving him cigarettes, and letting him have the horses he wants to work with.

"I can't whip him down," he says. "He's already killed people like me."

Finally, though, Buchanan sees to it that the aggressive man finds himself in a corral with the meanest, wildest bronc on the place—and an audience to watch the show. The outcome is a foregone conclusion, for even the toughest inmate is no match for a truly wild horse. After it's over, Buchanan moves in on the man.

"I beat him down," he says, "make him cry like a baby. Then he starts asking how to do things. After that he makes a really good hand."

In spite of this hard-nosed approach, Buchanan's program is a popular one here.

"Everybody likes horses," he says.

We drive back to the administration building and visit the farm superintendent, Gary Starbuck, in his office.

Gary Leppart

"What Mike is doing is a real art," he tells me. "He's the best psychologist I've ever met."

Starbuck says the wild horse program, which will soon celebrate its 10th anniversary at the farm, teaches the inmates patience and cognitive skills.

Buchanan sees the program as the difference between being "locked down" and "doing time." Being locked down, he says, is what happens in Rawlins, where "I do it all for you." Doing time, however, is participating in your own rehabilitation, learning to take responsibility for your life.

"We're doing the exact same thing getting a wild horse ready for society as we are with the inmate," he says.

Although trained horses are, from the standpoint of prison officials, merely a fortunate byproduct of this program, BLM spokesman Ron Hall is also enthusiastic about it. He says every one of these horses gets adopted. He is also impressed by Buchanan's ability to get the inmates to bond with the animals.

"For some of them," he says of these men, "it's the first time that's ever happened in their life."

As I pull out of the farm's parking lot, the metal corrals full of horses are already out of sight, but I can still see the chain link enclosure for the men. The two sets of prisoners here, I think, are each teaching the other. The horses are learning to survive in captivity, the men are learning to be free. ■

2 SOURCES AND SELECTED READINGS

Lynne Bama

Clutton-Brock, J. 1992. *Horse Power: A History of the Horse and the Donkey in Human Societies*. Cambridge, MA: Harvard University Press.

Lawrence, E. A. 1985. *Hoofbeats and Society: Studies of Human-Horse Interactions*. Bloomington, IN: University of Indiana Press.

About the Author

A Pennsylvania native, Lynne Bama graduated from New York University and the Germaine School of Photography before moving to Wyoming in 1968. Her writing has appeared in a number of publications, including *Sierra, Orion*, and *High Country News*. Lynne has written about wild horses for *Sierra Club Books* and *High Country News*. She lives in northwestern Wyoming with her husband, the artist James Bama, and their son.

John Eastcott and Yva Momatiuk

3 The Horse Nations Endure

Dan Flores

The sun's horse is a yellow stallion, a blue stallion a black stallion; the sun's horse has come out to us.

— Apache Ceremonial Song

Above: *Untitled,* the Horse in ledger art. Buffalo Bill Historical Center, Cody, WY.

Left: Gary Leppart

One warm autumn afternoon in 1997, Gerard Baker, then superintendent of Little Bighorn Battlefield National Monument, told me a story that has stuck. Gesturing at distant ridges across the vast plains east of the battle site, whence Custer had reconnoitered the field, Gerard remarked that the reason the Army's Indian scouts knew that a *big* village lay hidden in the valley was because in their glasses they could see, through shimmering heat waves, a hovering, roiling insect cloud. Custer, famously, could see no evidence of a camp. But the Indian scouts knew: an insect cloud that size meant a huge pony herd—a kind of knowledge bred from long, intimate, and rich association with horses.

With the possible exception of the mounted cowboy maneuvering a herd of steers, in modern world history there has never been a visual image of humans and animals that is so universally recognized—or so deeply resonant in the modern consciousness—as the one between American Indians, their horses, and bison. Western art has seen to that. From the moment in 1820 when Titian Peale of the Long Expedition first painted a Kiowa hunter, bow drawn to full tension,

Gary Leppart

kneeing his buffalo runner close in on a panicked buffalo, right down to the dramatic buffalo running scenes of Kevin Costner's 1990 film, *Dances With Wolves*, the world has internalized the Plains Indian, his brightly attired mount, and their shaggy prey as a set-piece of Americana.

The real wonder is not how, or even why, that particular relationship came to stand as such a potent symbol of the West and the wild, but what a rich historical tapestry this single image actually rests upon. The bison, for its part, has a past that stretches back 100,000 years in North America, through at least four species. The ancestors of the native peoples may have come as early as 25,000 years ago. But the horse? In its earliest form (the little striped *Hyracotherium*) the horse emerged out of the mammal radiation that followed a huge asteroid hit 65 million years ago. Horse taxonomists like Gaylord Simpson, Bruce McFadden, Richard Hulbert and others have pursued that ramification to the ground, tracing (and noting many dead-ends and extinctions along the way) what started as a sort of three-toed rabbit down to the modern family, *Equidae*. And the argument now is that for the last 15 million years right down to 10,000 years ago, horses may have dominated the fauna of North America, at times comprising a *third* (by number, if not biomass) of the big animals of western America. Depending on how the fossils get divvied up, at least *sixty* species of the genus *Equus* have been named from North America.

Those species include several whose time-spans obviously overlapped with the arrival from Siberia of the ancestors of the American Indians. So the most mysterious of all the interactions between horses and Indians is this first one. How did humans react to the teeming horse herds of the Americas? And what actually happened to the *Equus* herds that were still on the plains of the West until about 8,000 years ago?

The answer, quite astonishingly, is that while horses that migrated into Eurasia survived, in their home of native origin horses quite suddenly died out. Paleontologists still don't know quite what to make of this—particularly given the horses' apparent joyous response when Europeans returned them to the Americas thousands of years later—or what role the Clovis and Folsom hunters played in it. To most of us the argument that the early Indian cultures were responsible for this holocaust seems more than a little farfetched. Horse remains *are* found in Paleoindian kill sites, but "horse jumps" analogous to the horse slaughtering grounds of Solutre, France (where perhaps 100,000 horses were killed during the Pleistocene) aren't in the American record. Scholars of horse evolution may satisfy themselves by referring to the whole range of Pleistocene changes—climatic and vegetative as well as prehistoric overkill—but one suspects that with horses something big is left out of the Pleistocene extinction story.

This is a problem made even more puzzling by what transpired 75 centuries later, between 1494 and 1519, when horses were returned to the Americas from the Old World. The horses that now danced beneath Spanish *conquistadores* onto American soil were not quite the same animals that had vanished so mysteriously in the Pleistocene. Horses had continued to evolve in Eurasia, and the Barb stock that made up the bulk of Spanish mustangs had for centuries been bred in the Islamic world for the desert-like conditions around the Mediterranean. So the horse "fit" back into the Americas was not an exact return of a species to its long-vacant niche. The fit was apparently close enough, however, and the end result was like the opening of a floodgate. In a fraction of an ecological instant, those pre-adapted desert horses went kicking, galloping, and whinnying their way back into continental ecosystems. It was in every way a remarkable homecoming.

The way horses diffused from tribe-to-tribe across the West of North America is a well-known story by now, although new elements of the early stages of it are still coming to light. There seem to have been at least four different seminal seedbeds of horse diffusion in the West: Northern Mexico, where large ranches provided abundant stock, and accounts of mounted Indians appear as early as 1660; New Mexico after the Pueblo Revolt of 1680 liberated at least 2,000 Spanish horses, the surplus of which was traded to groups like the Utes and Navajos; East Texas, where

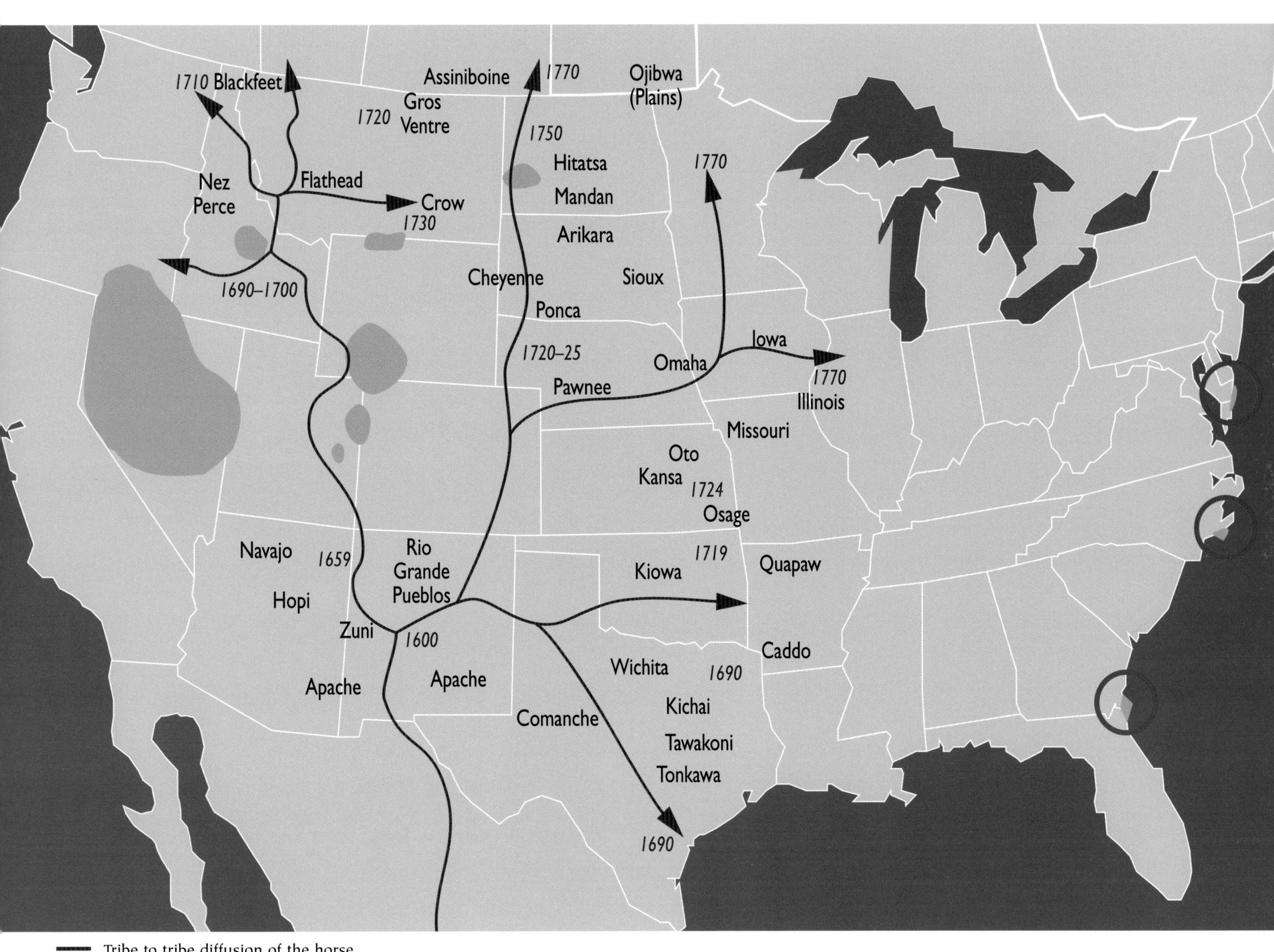
1710 Blackfeet
Assiniboine
1770
Ojibwa (Plains)
1720 Gros Ventre
1750
Hitatsa
Mandan
1770
Nez Perce
Flathead
Crow
1730
Arikara
1690–1700
Cheyenne
Sioux
Ponca
1720–25
Omaha
Iowa
1770
Illinois
Pawnee
Missouri
Oto
Kansa
1724
Osage
Navajo
1659
Rio Grande Pueblos
1719
Quapaw
Kiowa
Hopi
Zuni
1600
Caddo
Wichita
1690
Apache
Apache
Kichai
Comanche
Tawakoni
Tonkawa
1690
Tribe to tribe diffusion of the horse
Wild Horse ranges in the United States, 1999

Albert Bierstadt (1830-1902), *Last of the Buffalo*, c. 1888. Oil on canvas, $60^{1}/_{4}$ x $96^{1}/_{2}$ in. Buffalo Bill Historical Center, Cody, WY. Gertrude Vanderbilt Whitney Trust Fund Purchase.

animals abandoned around the short-lived missions of the 1690s increased to thousands, covering the whole countryside, by the time Spaniards returned in 1715; and the Southern California missions, where horses became so numerous during the 18th century that policies of extermination had to be waged to protect pasturage.

Exemplifying the principle that truly significant new developments in human affairs move with amazing rapidity from group-to-group, horses—and the cultural package of how to utilize and care for them— spread from these four scattered centers of diffusion to every Western setting that could support horses, and did so in the span of a single human generation. The best-documented transfers took place up the west slope of the Rockies, where Utes who acquired horses after the Pueblo Revolt quickly traded them northward to Shoshones and Bannocks in present Idaho and Wyoming . . . who took them to the Bitterroot Valley Salish . . . who then traded them northward and westward to groups like the Kootenais, the Nez Perce, and other mountain tribes. By 1730, only 50 years after the Pueblo Revolt, peoples as remote from New Mexico as the Blackfeet and Assiniboines of Alberta had horses, and the Black Hills was emerging as a kind of horse-trading center on the Northern Plains. A century later, when Lewis and Clark arrived in today's western Montana, they found groups like the Salish not only splendidly mounted, but obviously Iberian in their approach to tack, handling their herds, even gelding.

With Indian assistance the horse had returned, but the horses had a hand in their destiny. By 1800 some two million wild horses, or *cimarrons*, had overspread the Southern Plains, competing with bison for grass and water, and similar numbers were filling the valleys in California and west of the Divide in the Rockies. And not just returned; the horse was to bear revolution on its back. For pre-existing Plains groups, for gatherer-hunters west of the mountains, even for horticultural farmers planting their crops in the prairie river valleys, across much of the West horses had become a catalyst to transformation. Although certain elements of Indian societies, for example

Edward Borein (1872-1945), *Blackfoot Women Moving Camp*, No. 2, c. 1922. Etching and dry point on paper, 6 x 11 in. Buffalo Bill Historical Center, Cody, WY. Gift of Corliss C. and Audrienne H. Moseley.

priests—and women—among the Caddoan and Siouan farmers opposed this horse transformation, many Indians found the new lifestyle intoxicating, ripe with possibility. The consequences for history were enormous.

For roughly 250 years, then, during which these "Tartars" of the West captured the world's imagination, almost three-dozen different Indian "tribes" mounted up to become at least part-time buffalo hunters of the prairies. Because of escalating intertribal competition over the buffalo country, many of them—Utes, Shoshones, Nez Perces, and a host of others—lived and raised their horses in their mountain homelands and rode seasonally to buffalo, thus were "Plains Indians" only part-time. Mounted groups in the woodlands (some of the Cherokees and Choctaws) east of the Mississippi favored a similar strategy. Other Indians, however, virtually abandoned their former incarnations as gatherers or farmers, and during the whirlwind of continental change between 1750 and 1880 engaged in a form of ethnogenesis, or group re-creation. The rapidly evolving conditions of the bison hunt and the arrival of the global market to the West, along with tribal jockeying for territory, were emblematic of a new world. The Comanches, the Kiowas, the Cheyennes, the Crows, the Blackfeet, and the various Siouan divisions became the most well-known of these avatars of Indian life.

At the center of it all was the horse. Whether considered as beast of burden, life-altering technology, or new form of harnessed energy, horses were not mere intensifiers. They clearly worked a revolution on Indian cultures. In their earliest stages of horse acquisition, some Indian groups were so awed by the possibilities the animals offered that they were willing to trade their children for them. And when they got horses, they came to shape their entire existence around their herds, moving their camps to accommodate horse needs, sometimes retaining captives and slaves to use as herders, organizing acquisition and wealth—status, classes, and marriage—around them. Horses became such a form of wealth among Plains tribes that some (like Apauk, of the Blackfeet) complained: "Nowadays everyone is crazy about horses. . . . but why any man should want ten times as many horses as he can use, that is what I do not understand." Wealth and status were the reasons. Horses even shaped gender relations among many Indian groups, where a gender division of labor was the norm. Many kinds of gendered work became easier with horses, yet women among farming tribes often complained about horses in their gardens; one observer reported that the most common marital strife among the Pawnees was over horses. Horse acquisition became so central to the 19th century market economy revolving around buffalo robes that it even delayed marriages. But when an eligible man acquired horses his situation changed markedly, for a horse herd of 100-300 animals put him in a position to arrange multiple marriages. Horses thus encouraged male polygamy in many Indian societies.

The horse revolution didn't stop there. Plains Indian ethnobotany altered to become a medicine chest of what horses needed, and material culture (adapted from packsaddles, girths, and travois already in use with dogs) flowered, adding picket pins, puzzle hobbles, and the like to the Plains toolbox. A rich symbology of artistic horse decoration for tipis, bags, and shields sprang into being like some kind of stylized Plains Expressionism; indeed, where modern Plains Indian art began—on ledgers and among the Fort Marion, Florida, captives—horses were always the most-favored figures. Some famous horse Indians, like the Comanches, developed an entire taxonomy of different horse types. And some, like the Nez Perces, the Bitterroot Salish, possibly the Comanches, bred their animals for conformation, speed, color. Although some tribes would not do so, some Plains Indians (the Pawnees, the Comanches) came to rely on their horses as a lean-times food source, and so were in the process of evolving into true pastoralists, living off the animals they herded.

Horses influenced every aspect of the new Plains lifestyle, but most significant was their crucial role in new economies that brought Plains Indians into the global market. Across the Plains, from Texas to Canada, this took the form of the buffalo robe trade, which by the 1840s had reached a staggering volume made possible by well-trained "buffalo runners." Yet another horse-based economy for Indians, especially on the Southern Plains, was the direct exchange of horses to Anglo-American traders like Philip Nolan, Anthony Glass, and the Bent brothers. Tens of thousands of wild mustangs, Indian ponies, and horses stolen from Hispanic ranchers supplied the

Right: Joseph Henry Sharp (1859-1953), *Crow Indians on Horseback in Tipi Village*, black and white photograph, 11 x 13⅞ in. Buffalo Bill Historical Center, Cody, WY.

Alfred Jacob Miller (1810-1874), *Indians Fording A River*. Watercolor on paper, 4 3/4 x 6 in. Buffalo Bill Historical Center, Cody, WY. Gift of The Coe Foundation.

advancing American frontier in the 19th century. The horse trade further enmeshed groups like the Comanches, Wichitas, and Southern Cheyennes in the market, an influence that created wealth but did not bode well for the long term.

The Southern Cheyenne, one of the most skilled horse people of the Plains, had an axiom that, in part, explains the shaping role horses also came to have in intertribal strife. The best horse, the axiom went, is one you've raised as a colt and know; next best is a grown horse someone else has raised and trained but you have to learn; worst is a grown wild horse you've captured and whose idiosyncrasies are unknown. The axiom was not only an encouragement to raising horses, it became a goad for young men in search of status and wealth to engage in raids on other tribes for fine horses. For environmental reasons—winters took a dreadful toll on horses out on the Northern Plains, while pony herds in the mountain valleys and on the Southern Plains typically got through winters intact—Sioux, Assiniboine, and Blackfeet parties tended to raid southward and westward every spring, creating a ripple effect of intertribal antagonisms.

Honed by buffalo hunting (and people like the Comanches eventually hunted everything, from deer to wild turkeys, horseback), refined by horse raids and intertribal war, mounted Plains Indians became famous as perhaps the best light cavalry in the world. Yet while war leaders like Red Cloud and Crazy Horse were known for the tactics of their mounted warriors, their horse herds eventually made Plains Indians vulnerable. Weakened by nematode infestations and often inadequate winter nutrition, Indian horses ultimately rendered winter camps easy targets for U.S. troops during the Indian wars, as at Sand Creek (1864) and the Battle of the Washita (1868). Savvy commanders like Colonel Ranald Mackenzie eventually found that they could break Indian resistance by slaughtering captured pony herds, as Mackenzie did with more than 1,000 horses after the Battle of Palo Duro Canyon (1874).

Horses may have been a liability in the final clash with the American military, yet such was

Gary Leppart

their psychological and historical legacy that even as the old life ended for Indians, horses never lost their power. So potent and appreciated was their effect on Indian life that horse medicine, and the horse's role in ceremonial life (as in the detailed horse ceremonies of Black Elk's vision in *Black Elk Speaks*), had become fundamental to many western Indian cultures. The horse, we ought to remember, was the only European-derived element that the Ghost Dance of the late 19th century sought to retain in the new Indian millennium Wovoka prophesied.

So even as federal policymakers in the assimilation/acculturation period labored to wean Indians on western reservations of their "inordinate fondness" for keeping horses, the horse's power over the Indian imagination did not die. As if in fulfillment of prophesy, a century after the Ghost Dance both buffalo and horses are emerging as the symbols and very real agents of a cultural renaissance among many modern Western Indians. The Inter-Tribal Bison Cooperative, which during the 1990s returned buffalo to more than three-dozen tribes across the West, has served as a kind of model for those hoping to do the same with horses.

Among the tribes using horses to fuel their cultural renaissance, Montana's Blackfeet and the Nez Perce of Idaho have outstripped all the others. The Nez Perce tribe has been the pioneer, in 1991 assembling 13 Appaloosas (now grown to a herd of almost 50) on tribal land to help rejoin tribal members and their legendary animals. Through the programs of the tribe's Chief Joseph Foundation, which is making horses available to families that lack pasturage and is working to build an equine center on the reservation, Appaloosa horses have become essential players in linking tribal elders, history and young people.

Similarly, in 1996 the Blackfeet took delivery of eight registered mustangs acquired from famed Wyoming mustang breeder Emmet Brislawn, and the Blackfeet Buffalo Horse Coalition was born. By 1999 the Blackfeet mustang herd is up to 100 animals. Through programs involving both adults and children, the effect of these horses on the tribe's sense of its history and culture, according to Bob Blackbull, has been profound.

All of which is testimony that the kind of deep knowledge of horses that was second nature to Custer's Crow and Arikara scouts is not going to fade into mere history. The Horse Nations endure. ■

The Horse Nations Endure

Gary Leppart

SOURCES AND SELECTED READINGS

Dan Flores

Binnema, T. 1998. "Common and Contested Ground: A History of the Northwestern Plains from A.D. 200 to 1806." Ph.D. Dissertation. Edmonton, Alberta, Canada: The University of Alberta.

Black Elk. 1971. *Black Elk Speaks: Being the Life and Story of a Holy Man of the Oglala Sioux* as told through John G. Neihardt (Flaming Rainbow). Lincoln, NE: University of Nebraska Press.

Bozell, J. 1988. Changes in the role of the dog in protohistoric-historic Pawnee Culture. *Plains Anthropologist* 33:95-111.

Burkhardt, J. W. 1994. "Herbivory in the Intermountain West." In Burkhardt et al., eds. *Herbivory in the Intermountain West: an Overview of Evolutionary History, Historic Cultural Impacts and Lessons from the Past.* Walla Walla, WA: Interior Columbia Basin Ecosystem Management Project.

Clark, L. 1966. *They Sang for Their Horses: The Impact of the Horse on Navajo and Apache Folklore.* Tucson, AZ: University of Arizona Press.

Dobie, J. F. 1934. *The Mustangs.* New York: Bramhall House.

———. 1951, The Comanches and their horses. *Southwest Review* 36:99-103.

Duke, P. 1991. *Prints in Time.* Niwot, CO: University Press of Colorado.

Ewers, J. C. 1955. *The Horse in Blackfoot Culture.* Washington, D.C.: Smithsonian Institution.

Flores, D. In Press. "Where All the Pretty Horses Have Gone." In chapter 3 of *Horizontal Yellow: Nature and History in the Near Southwest.* Albuquerque: University of New Mexico Press.

———. 1991. Bison ecology and bison diplomacy: the southern plains from 1800 to 1850. *Journal of American History* 78:465-85.

Gelo, D. J. 1986. "Comanche Belief and Ritual." Ph.D. Dissertation. New Brunswick, NJ: Rutgers the State University of New Jersey.

———. 1990. "Comanche Horse Culture." *Thundering Hooves* exhibit planning sessions, Witte Museum, San Antonio, Texas, December 1990. [Notes in possession of author.]

Glass, A. 1985. *Journal of an Indian Trader: Anthony Glass and the Texas Trading Frontier,* 1790-1810. Edited by Dan Flores. College Station, TX: Texas A&M University Press.

Guthrie, R. D. 1980. Bison and man in North America. *Canadian Journal of Anthropology* (Revue canadienne d'anthropologie) 1:55-73.

Haines, F. 1938. The northward spread of horses among the Plains Indians. *American Anthropologist.* 40:429-37.

———. 1938. Where did the Indians get their horses? *American Anthropologist* 40:112-17.

———. 1963. *Appaloosa: The Spotted Horse in Art and History.* Austin, TX: University of Texas Press.

Hall, T. 1989. *Social Change in the Southwest,* 1350-1880. Lawrence, KS: University Press of Kansas.

Hanson, J. 1986. Adjustment and adaptation on the Northern Plains: the case of equestrianism among the Hidatsa. *Plains Anthropologist* 31:93-107.

Hines, J. 1998. The buffalo runners: hoofbeats across the centuries. *Montana Magazine* 152:24-29.

Holder, P. 1970. *The Hoe and the Horse on the Plains: A Study of Cultural Development among North American Indians.* Lincoln, NE: University of Nebraska Press.

Hulbert, R. Jr. 1977. "The ancestry of the horse." In *Horses Through Time,* edited by S.L.Olsen. Boulder, CO: Roberts Rinehart Publishers for Carnegie Museum of Natural History

Klein, L. and L. Ackerman, eds. 1995. *Women and Power in Native North America.* Tucson: University of Arizona Press.

McDonald, J. N. 1981. *North American Bison: Their Classification and Evolution*. Berkeley: University of California Press.

McFadden, B. J. 1992. *Fossil Horses: Systematics, Paleobiology, and Evolution of the Family Equidae.* Cambridge [England]; New York: Cambridge University Press.

McGinnis, A. 1990. *Counting Coup and Cutting Horses: Intertribal Warfare on the Northern Plains,* 1738-1889. Evergreen, CO: Cordillera Press.

Medicine Crow, J. 1992. *From the Heart of Crow Country: The Crow Indians' Own Stories.* New York: Orion Books.

Mishkin, B. 1940. *Rank and Warfare Among the Plains Indians.* Lincoln, NE: University of Nebraska Press.

Moore, J. 1987. *The Cheyenne Nation: A Social and Demographic History.* Lincoln, NE: University of Nebraska Press.

Olsen, S. 1997. "Horse Hunters of the Ice Age." In *Horses Through Time*, edited by Sandra L. Olsen. Boulder, CO: Roberts Rinehart Publishers for Carnegie Museum of Natural History.

Osburn, A. 1983. Ecological aspects of equestrian adaptations in aboriginal North America. *American Anthropologis*t 85:536-91.

Pinkham, A., President of the Chief Joseph Foundation, Lapwai, Idaho. Telephone interview with the author, January 26, 1999.

Roe, F. G. 1955. *The Indian and the Horse.* Norman, OK: University of Oklahoma Press.

Schultz, J. W. 1916. *Apauk: Caller of Buffalo.* Boston: Houghton-Mifflin.

Secoy, F. 1953. *Changing Military Patterns of the Great Plains Indians.* Lincoln, NE: University of Nebraska Press.

Sherow, J. 1992. Workings of the geodialectic: High Plains Indians and their horses in the Arkansas River Valley, 1800-1870. *Environmental Review* 16:61-84.

Simpson, G. G. 1951. *Horses: the Story of the Horse Family in the Modern World and through Sixty Million Years of History*. New York: Oxford University Press.

Vestal, P. and R. Schultes. 1939. *The Economic Botany of the Kiowa Indians*. Cambridge, MA: Botanical Museum of Harvard University.

West, E. 1998. *The Contested Plains: Indians, Goldseekers, and the Rush to Colorado*. Lawrence, KS: University Press of Kansas.

Wilson, C. 1963. An inquiry into the nature of Plains Indian cultural development. *American Anthropologist* 65:355-69.

About the Author

Dan L. Flores is A.B. Hammond Professor of Western History at the University of Montana, where he has taught since 1992. He has also held positions at Texas Tech University, the University of Wyoming, Texas A&M University, and Northwestern State University. He received a Ph.D. in History from Texas A&M University in 1978. Dr. Flores has written extensively on environmental history topics and has received numerous prizes for books and articles from such organizations as the Western History Association, the National Cowboy Hall of Fame, and Western Writers of America. He has two new works [*Horizontal Yellow* (University of New Mexico Press) and *The Natural West* (University of Oklahoma Press)] scheduled to appear in the fall of 1999 and spring 2000. Dr. Flores is also a regular contributor to *The Big Sky Journal*, *High Country News*, and *Montana, the Magazine of Western History*.

4

Freedom Versus Management: The Dilemma for Wild Horses

Linda Coates-Markle

Above and left: Gary Leppart

Truly "wild and free" equids disappeared from the North American continent almost 10,000 years ago, at least according to the fossil record. The Spanish conquistadors probably never intended to reintroduce vast herds of free-roaming horses on this continent, but that is exactly what happened. Almost 500 years ago, they imported horses from their homeland and established the first breeding farms in the West Indies, thus supporting a rapidly growing demand for horses. The use of horses spread throughout the continent, as barter and trade activities were popular, especially with the Native American tribes. Inevitably, many horses were released or stolen or escaped from their owners, and found opportunities to populate vast areas of the open range. Over the next several hundred years, additional domestic horses joined their ranks, and early Spanish characteristics of the horses became less evident. The term "feral," often used in reference to these wild herds, simply refers to the fact that these horses originated from once domestic stock. As a result, many of the herds which currently exist in the western United States contain diverse conformational traits reflecting obvious dilution by local domestic breeds.

While the majority of wild populations currently reside in Nevada, several other states also provide habitat areas for specific herds. Some skeptics are concerned that the reappearance of horses, as exotics, on western rangelands will challenge the ability of these ecosystems to readapt to their grazing pressure. Others believe that the systems will readapt as readily as the horses have, and are very much in favor of conservation activities to ensure long-term survival of the populations. Of particular interest are a few rare situations where it is believed that significant isolation from domestic horses occurred during the development of these herds. Interestingly, these horses tend to bear early Spanish-type traits in conformation and color. Research has shown that some of these populations are also genetically unique, possessing gene material not found in many of today's domestic breeds. One such herd resides within the Pryor Mountain Wild Horse Range on the border of Montana and Wyoming. This range was established in 1968 in an effort, driven by local and national public concern, to conserve the perceived uniqueness of this herd.

Behaviorally, though, noticeable similarities exist between most wild horse populations in the western states, with harems and bachelor bands particularly evident. Dominant males generally acquire and defend females within groups of four to six, and this family remains intact, as a unit, while traveling to new areas in search of available forage and water. This type of social structure facilitates the nomadic tendencies of these animals and probably promoted their historical spread and rapid adaptation to new environments. In the past, if an area was unsuitable for habitation, the animals simply moved on and continued exploring new areas. This active dispersal probably continued until the rangelands were subjected to extensive fencing, in the 1800s, in response to livestock interests. During this time, hundreds of thousands of wild horses were captured and removed from the range for war efforts, the agricultural industry and human and pet food. Eventually, however, public concern over methods of horse capture and transport resulted in the passing of one of the more controversial laws in American history.

Velma Johnston, better known as Wild Horse Annie by her critics, was one of the more concerned citizens instrumental in channeling public outcry and efforts in support of legal protection for the horses. As a result of her early efforts, a 1959 bill named in her honor was passed by Congress, which made it illegal to gather the horses using motor vehicles or airplanes. Continued diligence on her part finally resulted in passage of the Wild Free-Roaming Horse and Burro Act in 1971. This law affords all wild equines on public lands protection and humane management by the secretaries of Interior and Agriculture. As such, Congress finds and declares that wild free-roaming horses and burros are living symbols of the historic and pioneer spirit of the West. Stipulations within the Act specifically state that recognized herds must have existed within a known area at the time of the Act. Subsequent management often requires that artificial barriers be constructed to further define the limits of designated herd areas, and prevent animal travel outside of these areas. What is most important, however, is that the Act specifies that herds be managed, at a

Gary Leppart

John Eastcott and
Yva Momatiuk

minimum feasible level, as "self-sustaining populations," and within the framework of a "thriving natural ecological balance" on the public lands.

Unfortunately, almost 30 years later, appropriate definitions of these concepts still remain subjective and somewhat elusive to managers. Much controversy still surrounds the determination of appropriate numbers of horses (or burros) within each designated area. Supporters of the horses are concerned that sustainable populations, promoting long-term genetic viability, fail to exist when population numbers fall below 100 individuals. Although this occasionally happens in many herd areas, often much mixing occurs between populations such that smaller herds are really subsets of much larger populations. Conversely, supporters of other approved uses of public lands generally have concerns that horse populations are too large for most areas and will eventually cause range degradation, or are consuming valuable range resources better designated for livestock production or wilderness values. In the end, a true balance will be reached only with the careful consideration and compromise of all legitimate uses of the rangelands.

Sound management involving the horses requires an understanding of the inherent processes that contribute to a balanced system in nature. Generally this information comes from intensive study of populations within their natural settings. For example, in a remote wilderness near the California-Nevada border, a herd of wild horses maintains its population size despite heavy foal mortality due to mountain lions. Horses within this herd have developed behavioral patterns promoting their survival in the face of this aggressive predation. However, within the nearly 200 designated Herd Management Areas found in the western United States, this scenario is not the norm. It is speculated that only two to three other herd areas may be vulnerable to significant impacts of natural predation, while the remaining herds are typically subjected to human-induced means of population control. This dilemma and its associated controversy, surrounding both the methods and necessity for population control, form the core of the Bureau of Land Management's

Wild Horse and Burro program.

Monitoring and research have shown that many wild horse herds currently enjoy substantial growth rates approaching an 18-25 percent increase in population size each year. While other wild ungulate species, like deer and elk, become limited by available resources as their populations grow, herds of horses are generally maintained below what is referred to as a "food-limited" density. This means that population size does not fluctuate around a naturally induced optimum, but continues, between human-induced reductions, to grow with minimal negative consequences. The result, in management terms, is a population producing a "maximum sustained yield" due to maximum birth and survival rates each year. If a horse population were to eventually reach its food-limited density, the result would first be visible in the overall condition of the herd. Individual animals would show poor health and fitness, foal production would decline, and adult mortality might increase through disease or vulnerability to predation. In time, the supporting range would show signs of overuse and the annual production of forage would decline. In fragile arid environments (as found in many parts of the western states), the range may take years to recover from this overexposure to grazing and additional trampling impacts.

Management objectives, then, strive to keep both the herds and range healthy by promoting active foal production and struggling to determine the appropriate population density for a given range area.

Right: John Eastcott and Yva Momatiuk

Because of the general lack of natural predation, management must mimic this activity, usually in the form of population gathers with an ultimate reduction in population size. Animals are generally removed selectively, and then placed in "pre-approved" homes through the Bureau's Adopt-A-Horse program. Adoption is currently the best tool the Bureau has for providing humane care of these animals subsequent to removal from the range. Throughout this whole process, an astute manager tries to sustain the basic integrity of the natural population, minimize intrusive management, and maintain the "fittest" animals on the range. Generally this requires that a manager understand the basic age structure, sex ratio and social configuration of the population, while resisting temptations for anthropocentric analogies. In this manner, the core of the population is maintained, and free to determine processes pertaining to social interactions, breeding success and survival. Fortunately then, elements like natural selection will have an opportunity to impact the surviving gene pool of the population. This is absolutely critical because long-term population viability is determined by the health of this gene pool. If too manipulative, human managers could, unknowingly, create serious long-term negative impacts on the population.

Smaller populations may be particularly vulnerable to these types of management manipulations. Much effort is being focused on the determination of minimum viable (self-sustaining) population sizes under different management scenarios. Viable numbers will permit long-term survival of populations. Researchers are looking at ways to model these populations, within the context of their environment, to maintain viable numbers while subjecting populations to various types of control. Controls may involve removal of select individuals of different ages or sex, application of birth control or a combination of both techniques. Recent developments in "immunocontraception" have gained support from a growing sector of the concerned public. Within this approach, targeted mares receive a vaccine which prevents the ability of the sperm to penetrate and fertilize the egg. As a result, the mare will ovulate and more importantly, behave normally, but

Dewey Vanderhoff

will not become pregnant. Used judiciously, and under the auspices of good science, this technique can help with population control without subjecting horses to the intrusion of gathers and the stress of relocation within the adoption program. As with all new technology, however, much research is needed prior to application.

The challenge to management is to mimic natural ecological processes which affect population birth and death rates, while maintaining viable populations and minimizing long-term deleterious impacts on horse herds and their surrounding environment. To be successful in this arena is no easy feat. The biggest problem is that much of nature is stochastic, or unpredictable, from day to day (even minute to minute), and one can never really hope to manage the ecosystem. We can only hope to manage within the context of the ecosystem—or in other words, within acceptable limits. This approach is referred to as "adaptive resource management," where decisions are made on the best possible information available, and the effects of decisions are constantly monitored and revised for improvement. As such, ecosystems and their components may be modeled in an effort to forecast future events. This requires intensive monitoring of multiple levels of all components of the system, from the soils to the variable climate. The horse populations then become only one part of a much larger and multifaceted system, and astute management becomes more than just a singular focus on horse numbers and estimated forage production within a defined range area.

Development of this level of management effectiveness will take some doing and will involve cooperation and collaboration between agency personnel responsible for management, the concerned public and the researchers who are developing these techniques. Predictive modeling has definite application in management, but models are only as good as the variables that define them and the accuracy of the selected data. Thus it becomes even more important to gain an understanding of the attributes that define individual populations in terms of age, sex and social

Right: John Eastcott and Yva Momatiuk

structure and how these attributes interact with other environmental variables to drive natural oscillations of population size. The presence and impact of predation, disease and possible competitive interactions with other animal populations must be evaluated within the context of these models. Only then, despite limitations imposed by management on population size and available range, can we hope that the determinations of survival are left primarily to the individuals of the species. In this manner, the horse populations will continue to remain as "free" as possible within a diminishing world. ■

Gary Leppart

4 SOURCES AND SELECTED READINGS

Linda Coates-Markle

Berger, J. 1986. *Wild Horses of the Great Basin: Social Competition and Population Size.* Chicago: University of Chicago Press.

Boyles, J.S. 1986. Managing America's wild horses and burros. *Journal of Equine Veterinary Science* 6(5):261-265.

Coughenour, M. B. 1991. Spatial components of plant-herbivore interactions in pastoral, ranching, and native ungulate ecosystems. *Journal of Range Management* 44(6):530-541.

Duncan, P. B. 1992. *Horses and Grasses: The Nutritional Ecology of Equids and Their Impact on The Camargue.* Ecological Studies 87. New York: Springer Verlag.

Ellis, J. and M. B. Coughenour. 1998. "The SAVANNA Integrated Modelling System: An Integrated Remote Sensing, GIS and Spatial Stimulation Modelling Approach." Chapter 7 of *Drylands: Sustainable Use of Rangelands Into the Twenty-First Century*, edited by V. R. Squires and A. E. Sidahmed. IFAD Series, Technical Reports. Rome, Italy: IFAD.

Fazio, P. M. 1997. The fight to save a memory: creation of the Pryor Mountain Wild Horse Range. *The Wyoming History Journal* 69(2):28-47.

Garrott, R. A. 1990. "Demography of Feral Horse Populations in the Western United States." Ph.D. Dissertation. Minneapolis: University of Minnesota.

———. D. B. Siniff, J. R. Tester, T. C. Eagle, and E. D. Plotka. 1992. A comparison of contraceptive technologies for feral horse management. *Wildlife Society Bulletin* 20:318-326.

Goodloe, R. B., R. J. Warren, E. Gus Cothran, S. P. Bratton, and K. A. Trembicki. 1991. Genetic variation and its management applications in eastern U.S. feral horses. *Journal of Wildlife Management* 55:412-421.

Huffaker, R. G., J. E. Wilen, and B. D. Gardner. 1990. A bioeconomic livestock/wild horse trade-off mechanism for conserving public rangeland vegetation. *Western Journal of Agricultural Economics* 15:73-82.

Kirkpatrick, J. F. 1995. Management of Wild Horses By Fertility Control: The Assateague Experience. *Scientific Monograph*, 92/26. Denver, CO: United States Department of the Interior; National Park Service.

Pitt, K. P. 1985. The wild free-roaming horses and burros act: a western melodrama. *Environmental Law* 15(3):503-531.

Rubenstein, D. I. 1986. "Ecology and Sociology in Horses and Zebras." In *Ecological Aspects of Social Evolution*, edited by D. I. Rubenstein and R. W. Wrangham. Princeton, NJ: Princeton University Press.

Thomas, H. Smith. 1979. *The Wild Horse Controversy*. S. Brunswick, NJ: A. S. Barnes.

Turner, J. W., Jr., M. L. Wolfe and J. F. Kirkpatrick. 1992. Seasonal mountain lion predation on feral horse population. *Canadian Journal of Zoology* 70:929-934.

———. I. K. M. Liu, A. T. Rutberg and J. F. Kirkpatrick. 1997. Immunocontraception limits foal production in free-roaming feral horses in Nevada. *The Journal of Wildlife Management* 61(3):873.

United States. Bureau of Land Management. 1997. The 10th and 11th Report to Congress on the Administration of the Wild Free-Roaming Horse and Burro Act for Fiscal Years 1992-1995. Washington, D. C.: U. S. Department of the Interior, Bureau of Land Management: U.S. Department of Agriculture, Forest Service.

United States. General Accounting Office. 1990. Rangeland Management: Improvements Needed in Federal Wild Horse Program: Report to the Secretary of the Interior. Washington, D. C.: The Office. [GOV DOC NO: GA 1.13:RCED-90-110]

About the Author

Linda Coates-Markle is a wild horse and resource management specialist. Since 1995 she has worked for the Bureau of Land Management as the Montana/Dakotas Wild Horse and Burro Specialist. In this position, she oversees research and management efforts pertaining to the Pryor Mountain Wild Horse Range. Prior to this, Linda was employed as director of the Equine Sciences program at Oregon State University. She is also currently completing a doctoral degree focused on the ecological relationships between social structure and habitat use within a Nevada wild horse population.

The opinions expressed within this essay do not necessarily represent either the opinions, positions or perspective of the Bureau of Land Management.

Right: Gary Leppart

CMRussell
1915 ©

5 Wild Horses in Popular Culture

B. Byron Price

"The Mustang will go down through the years as one of the most colorful and charming figures in our history, and if perchance he at last becomes little more than a tradition, he will remain a delight and an inspiration to youthful imaginations, and retain a deathless affection in the hearts of Americans."

—William Robinson Leigh, *The Western Pony* (1933)

The romantic image of horses running wild and free in the untamed wilderness West has been an icon of American popular culture for nearly two centuries. Thanks to art, literature, film, music and advertising, wild horses have come to stand not only for the spirit of the West but also as the hope for its future. The meaning of such imagery, however, has varied from generation to generation. To early chroniclers, wild horses were a noble part of an exotic and untainted natural landscape, though subject to the taming impulses of civilization. Today the descendants of these mustangs represent an obstacle to progress to some and to others the admirable outcasts of civilized society, anti-heroes worthy of governmental protection in their "natural" state.

Whatever the time, wild horses have represented a freedom from restraint that has always appealed to Americans. As the plight and fate of real mustangs grow more precarious and as humans face an increasingly mechanized, urban and impersonal existence, this "wild and free"

Left: C. M. Russell, (1864-1926) *Broncho Buster* (detail). Watercolor on paper, 11 x 16⅝ in. Buffalo Bill Historical Center, Cody, WY. Gift of William E. Weiss.

imagery becomes ever more compelling. Savvy advertisers have taken full advantage of these realities and the anxieties they produce and have carefully packaged the image of untamed, defiant wild horses to sell everything from automobiles to wine.

That wild or feral horses endure at all in today's world is something of a miracle, given the extraordinary efforts made to destroy them. Yet their vibrant image on printed page, on painted canvas, in solid bronze, in lilting lyrics and in the popular imagination has undoubtedly contributed to their survival. Not surprisingly, congressional legislation passed in 1971 protecting wild horses called them, "Living symbols of the historic and pioneer spirit of the West."

The word "mustang" entered the English lexicon from the Spanish word *mesteño*, as Anglo-Americans began to encounter vast herds of feral horses on the Great Plains in the early nineteenth century. Tall tales told around the campfire by "mustangers," who chased wild horses for sport and profit, spread the lore of these colorful creatures from the northern Rockies to the Rio Grande.

The most prominent and enduring of all these frontier yarns is that of the "Pacing White Stallion," or "White Steed of the Prairies," a mustang with such bearing, speed and power as to inspire the awe and envy of all who tried to capture it. Texas folklorist J. Frank Dobie, who tracked the origins and development of this tale across 19th century frontier America, credits Washington Irving with the first published version in the 1832 travelogue *A Tour of the Prairies*. Over time, other writers repeated and embellished the story of the white pacing stallion in books, poems and plays. Although each retelling celebrated the heroism and intelligence of the wild stallion, few paid him more effusive homage than Herman Melville's classic seafaring tale, *Moby Dick*. When comparing the great whale's whiteness to that of other notable white objects, the author recalled the

. . . White Steed of the Prairies; a magnificent milk-white charger, large-eyed, small-headed,

Alfred Jacob Miller (1810-1874), *Wild Horses*, c. 1837. Wash drawing on blue paper, $8^{1}/_{2}$ x $13^{3}/_{4}$ in. Buffalo Bill Historical Center, Cody, WY. Gift of The Coe Foundation.

George Catlin (1796-1872), *Catching the Wild Horses*, lithograph, hand colored, 13 x 17$^1/_2$ in. Buffalo Bill Historical Center, Cody, WY. Gift of Mrs. Sidney T. Miller.

bluff-chested, and with the dignity of a thousand monarchs in his lofty, over-scorning carriage. He was the elected Xerxes of vast herds of wild horses, whose pastures in those days were only fenced by the Rocky Mountains and the Alleghenies. At their flaming head he westward trooped it like the chosen star which each evening leads on the hosts of light. The flashing cascade of his mane, the curving comet of his tail, invested him with housings more resplendent than gold-and silver-beaters could have furnished him. A most imperial and archangelical apparition of that unfallen, western world, which to the eyes of the old trappers and hunters revived the glories of . . . primeval times—in whatever aspect he presented himself, always to the bravest Indians he was the object of trembling reverence and awe.

No wonder showman P.T. Barnum is said to have offered a reward of five thousand dollars to whomever captured the animal!

In most versions of the tale, however, the great white stallion eludes his captors and relocates to some distant range. In a few variations the mustang dies, usually by accident, natural predator, or a self-determined act like starvation. Only occasionally is he captured or killed by humans.

Tales of an invincible white mustang not only fueled literary imaginations, but also inspired artists, whose depictions of wild horses reflected upon the twin themes of freedom and subjugation. George Catlin, probably the first Euro-American to paint free-ranging western mustangs, depicted wild horse herds frolicking in the wilderness or being chased and subdued by Native Americans on the Great Plains. In Catlin's works, the regal bearing of tame mustangs ridden by resplendent Indian warriors contrasts markedly with that of their scruffy four-legged cousins who still roamed the prairie.

About the same time Catlin encountered wild horses on the Plains, that careful observer of nature, John J. Audubon, bought a freshly tamed mustang in Kentucky. "The animal," wrote the artist, "was by no means handsome; he had a large head, with considerable prominence in its frontal region, his thick and unkempt mane hung along his neck to the breast, and his tail, too scanty to be called flowing, almost reached the ground. But his chest was broad, his legs clean and sinewy, and his eyes and nostrils indicated spirit, vigor, and endurance." Despite these admirable qualities, Audubon did not include wild horses in his famous artistic renderings of western quadrupeds.

Artists in the field rarely got more than a distant glimpse of the herds of mustangs that came within their view. Painter Alfred Jacob Miller, who accompanied an expedition to the Rocky Mountains of Wyoming in 1837, usually settled for telescopic observation. But even at a distance, the sight of a free roaming horse herd left a profound impression on the young Baltimore artist. "Among the wild animals of the West," he waxed,

> "—none gave us so much pleasure or caused such excitement, as the bands of wild horses that at intervals came under our view:—The beauty & symmetry of their forms,—their wild & spirited action,—long full sweeping manes & tails, variety of colour, & fleetness of motion,—all combined to call forth admiration from the most stoical:—One of the greatest difficulties we experienced was to get near enough:—They fought shy & held us at a long range,—showing that they were prudent & sensible in addition to other fine qualities.—Often we had to resort to a telescope.—They wheel like trained columns of cavalry,—charge, scatter, & form again.—Anon they are seen in battalions scampering across the prairie, stopping for a moment,—snuffing the breeze,—taking a final look at the intruders from the last undulation,—and are gone.

The sketch will convey to the observer some idea of this glorious scene,—but it is almost impossible to catch such magic convolutions, & secure the spirit of such evanescent forms,—under the excitement and difficulties that may be readily imagined to transpire at the moment.

Miller's renderings of wild horses convey a romantic view that matches his verbal descriptions and owes much to prevailing artistic conventions. [For typical examples of Miller's wild horse sketches see *Herd of Wild Horses* and *Wild Horses,* Whitney Gallery of Western Art, Buffalo Bill Historical

Center, Cody, Wyoming, and *Stampede of Wild Horses*, Gilcrease Museum, Tulsa, Oklahoma.] As art historian Joan Tricolli has observed: "Miller's horses, the very embodiment of Romantic freedom, belong to a noble breed more closely related to the Arabian stock painted by European masters like Eugène Delacroix than the wild ponies of the Great Plains."

A year before Miller discovered wild horses on the northern plains, William T. Ranney, a New York art student then serving in the Army of the Republic of Texas, was sketching them on the salt grass prairies of the Gulf Coast. A decade later he produced *Hunting Wild Horses*, a dramatic painting that depicted the often-ferocious battles between mustangers and their unwilling quarry. In this classic rendering, a lasso-wielding harbinger of civilization subdues a white stallion, the embodiment of nature.

The theme of subjugation continued to flourish in the late 19th and early 20th century illustrations, easel paintings and sculpture of artists like Frederic Remington and Charles M. Russell. These artists, however, infused their narratives with realism heretofore lacking in earlier depictions of the subject. Through such widely reproduced works as Russell's *Wild Horse Hunters* and Remington's *His First Lesson* and *The Bronco Buster*, among others, the capture and taming of wild horses served as a powerful metaphor for the taming of the West itself. [See Charles M. Russell, *Wild Horse Hunters* and Frederic S. Remington, *His First Lesson* and *The Bronco Buster*, all in the collection of the Amon Carter Museum, Fort Worth, Texas. Frederic Remington's and Charles M. Russell's artistic interpretation of the American West are the subject of countless books and articles. Some of the best include Alex Nemrov, *Frederic Remington and Turn-of-the-Century America* (New Haven: Yale University Press, 1995); Michael E. Shapiro et. al., *Frederic Remington: The Masterworks* (New York: Harry N. Abrams Inc., 1988); Brian W. Dippie, *Remington & Russell* (Austin: University of Texas Press, 1982); Brian W. Dippie, *Looking at Russell* (Fort Worth: Amon Carter Museum, 1987).]

While American artists seemed obsessed with domesticating wild horses, French animal painter Rosa Bonheur celebrated their continued freedom. Although Bonheur never came West to paint her subject in the field, the gift of a pair of wild horses in the mid-1880s afforded her an opportunity to study the real thing first-hand. The *chevaux sauvages*, one called *Apache* and the other with the ill-fitting moniker *Clair-de-Lune*, were a gift of American coffee magnate and Wyoming rancher John Arbuckle. The animals eventually proved too troublesome for the artist to keep, however, and Buffalo Bill Cody, whose famous Wild West came to Paris in the fall of 1889, took them off her hands.

The bright-colored posters that heralded Cody's show in cities across America and Europe often featured the bucking antics of the *remuda* of mustangs that regularly performed with the troupe. In one of the most picturesque of these billboard-sized lithographs no less than a dozen determined cowboys attempted to master an equal number of swirling, pitching broncs while Cody looked on approvingly. The showman billed the melee as *Cowboy Fun*.

Rosa Bonheur frequented the Wild West show grounds during its seven-month run in the French capital and converted her observations into countless sketches and several paintings of the Indians, buffalo and other animals. Toward the end of her life, Bonheur embarked on an ambitious painting with a wild horse theme. Desiring as much authenticity as possible in her canvas, *Race of Wild Horses*, Bonheur got an American friend to send her some western sagebrush and weeds, which she painted into the foreground. Unfortunately, the artist never completed the huge (50"x 87") work and it lay unfinished upon her easel at her death in 1899.

Of the several artists who found wild horses compelling subjects into the early 20th century, none surpassed William R. Leigh's painted portrayals. Leigh took his first trip West in 1906 and in the years that followed observed wild "broomtails" first-hand in the badlands near Cody, Wyoming, and in South Dakota and Arizona. From these experiences the artist absorbed much of wild horse lore and habits, which he interpreted in his drawings and paintings. Like Remington and

Right: George Catlin, *Wild Horses at Play in the Prairies of the Platte*, c. 1855-1870. Oil on paperboard on Bristol board, 18½ x 25 in. Buffalo Bill Historical Center, Cody, WY. Gift of Paul Mellon.

Russell, whose work influenced his own, Leigh was drawn to dramatic action and to epic life and death battles of men and animals, played out against a powerful western panorama. In Leigh's *The Killer*, for example, a cougar has dropped from a tree and clings tenaciously to the head of a mustang, which fights to dislodge it. A wild stallion defends his beleaguered band against a winter attack from wolves in another of Leigh's compositions, *The Struggle for Existence*. Although true to nature, the painting's design and mood resemble that of Frederic Remington's *Broncos and Timber Wolves*, painted in 1888 and published in Theodore Roosevelt's *Ranch Life and Hunting Trail*. Perhaps the finest of all Leigh's wild horse paintings is *The Leader's Downfall*. On this canvas muscular Indians mounted on muscular ponies rope muscular mustangs amid a muscular landscape. In this work, as in many others, Leigh exaggerated both the players and the action to accentuate both the drama of the scene and the personality of the characters.

William R. Leigh was not only an accomplished artist but also a fine writer. He devoted several chapters of his book *The Western Pony* to wild horses and their impact on western American life and culture. Of them he wrote admiringly: "There is no animal ranging the earth whose senses are more acute, who is more adroit, more resourceful or more courageous."

Mustangs found another 20th century champion in a cowboy named Will James. No one painted wild horses nor described them in print with more fidelity than this self-taught artist and plain-spoken writer. Beginning in the 1920s and continuing until his untimely death in 1942, James wrote and illustrated a steady stream of popular books and short stories, some of which remain in print today.

"Will James," said one admirer "truly loved horses all his life, and he loved the wild ones most of all." In works like *Smoky the Cowhorse, Cow Country, The Three Mustangers* and *Horses I've Known*, James brought the existence and the predicament of wild mustangs to the attention of thousands of readers. He wrote largely from experience and direct observation, his spirited narratives sometimes touching on the nagging dilemma of too many feral horses for too little range. As the youthful thrill of trapping mustangs waned with age, maturity and a prison term in Nevada, James took a more ambivalent view about depriving wild horses of their freedom. In a 1930 short story James mused: " . . . it wasn't owning the wild horse that made me want to go after him so much, it was the catching of him that caused a feller to get the mustang fever, and after the mustang was caught and the fever cooled down—well, I'd kinda wished they'd got away."

Filmmakers eventually adapted several of Will James' wild horse tales into motion pictures, including three screen versions of *Smoky the Cowhorse*, an animal with mustang lineage. These films, along with Zane Gray's novel *Wild Horse Mesa*, also filmed three times, joined a host of other motion pictures produced since 1920, whose story lines include mustangs. Movies with titles like *The White Outlaw, Wild Horse Canyon, Wild Horse Stampede, Tony Runs Wild,* and *Wild Beauty* and featuring cowboy stars like Jack Hoxie, Ken Maynard, Hoot Gibson and Tom Mix entertained a generation.

Many such films were aimed at children. Most were simple adventure stories, heavy on action and light on plot and character. *Konga, the Wild Stallion, Silver Stallion, King of the Wild Stallions* and *Last of the Wild Horses*, all produced in the 1940s, typified this escapist fare. The improbable scenario of the 1949 movie *The Golden Stallion*, for example, featured a tame palomino mare that traversed the Mexican border with a herd of wild mustangs, while carrying smuggled jewels hidden in a special horseshoe. This film starred Roy Rogers, Dale Evans and Roy's horse Trigger. Roy foiled the illicit gem traffickers, while Trigger courted the mare and trampled an abusive smuggler.

Camera-shy, real mustangs rarely appeared on film. Movie producers relied instead on generic horse herds with a well-trained animal or two on hand to represent fighting stallions or handle other stunt work. A horse called Rex, billed on theater marquees as *The King of the Wild Horses*, starred in several early films and was followed by several other mustang impersonators with names like "Tiger," "Starlight," "Thunder," and "Wildfire."

Until the 1961 release of *The Misfits,* starring Clark Gable, Marilyn Monroe and Montgomery Clift, wild horses were simply bit players in so many melodramas. Arthur Miller's screenplay, however, aspired to a realism and pointed social commentary heretofore absent in wild horse-related cinema. In Miller's script the flaws and fate of wild horses reflected metaphorically on the degraded condition of the movie's human characters.

As the film reaches its climax, Gable and Clift, two over-the-hill cowboys, mount a flat-bed truck to chase a tiny band of mustangs located by an airplane. The pair subdues their wild prey with ropes attached to automobile tires, intending to sell them as dog food. Monroe, a recent divorcee along for a lark, protests the cruelty, whereupon the cowboys, in danger themselves of being ground up by life, finally relent and turn the captives loose.

Montgomery Clift's portrayal of a tormented bronc rider in *The Misfits* also highlights rodeo's role in reinforcing the cultural and historical linkage between humans and wild horses in the West. Anthropologist Elizabeth A. Lawrence has noted that the performance and ritual inherent in rodeo events like bronc riding and wild horse races serve to reaffirm the pastoral values, lifestyle and traditions of westerners. The same may be said of the government-sponsored roundups held in many areas to thin wild horse herds. Through such events modern-day cowboys re-enact the conquest of nature while demonstrating the horseback skills that confirm "their place within the local social order."

Over the last century, the image of wild cowboys atop wild broncs has become a staple not only of western art and literature but also of western music. Songs about the "Zebra Dun" and the "Strawberry Roan," whose physical descriptions and rank disposition mark their mustang blood, are cowboy standards. Contemporary songwriters and performers like Michael Martin Murphey,

From the movie *The Misfits*, 1961 United Artists Corporation. © The Everett Collection.

Andy Wilkinson and Garth Brooks, among others, continue to tap both the reality and symbolism of wild horses to entertain and challenge modern audiences. Murphey's adaptation of a Blackfoot tribal legend of "The Ghost Horse," resulted in the haunting melody "Running Shadow." In "The Mustang Song," Wilkinson ties the ultimate fate of wild horses to the destiny of the West itself. Garth Brooks' "Wild Horses" strikes a more personal note when he sings of a rodeo cowboy whose existence seems as rootless as that of the broncs he rides.

The unbridled appeal of wild horses is not confined to country and western tunes but also resonates in the lyrics of several recent songs by classic rock groups. In one, the Rolling Stones assure listeners that wild horses cannot loosen a committed lover's tenacious grip on a relationship. In another, the Irish band U2 sings of a souring affair that leads one once-passionate paramour to ask the other, "Who's gonna ride your wild horses?"

The image of wild horses as free-spirited and nonconformist has so thoroughly permeated American popular culture in recent years that astute advertisers have found it an attractive means of selling products. Advertisements with wild horse themes typically beckon consumers to change their habits, to break away from the conventional and to experience a new kind of freedom. Other commercials trade on the swiftness and mobility of the breed. Surely the Ford Mustang automobile, a fixture on American highways since 1964, owes much of its enduring appeal to the mystique of its four-legged namesake. The same could be said of the broncos and mustangs that

1998 Ford Mustang. Photo by Renee Tafoya.

serve as popular sports mascots at every level of play in America. From the Wild Horse Winery of California to the Wild Horse Stampede Rodeo of Wolf Point, Montana, to Nashville's Wildhorse Saloon, haven for country music's line-dancing crowd, many popular businesses and events trade profitably on their association with the wild horses of the West.

Although the power and pervasiveness of wild horse imagery shows no sign of abating, its icons and vehicles of expression are continually evolving. In recent years, for example, photography, abetted by telephoto lenses and a host of glossy coffee table books, has supplanted the role that fine art and illustration once played in conveying the image of wild horses to Americans. Likewise the image of the bronc buster and the notion of "breaking" horses is giving way to the more humane methods espoused by horse trainers like Ray Hunt, the Dorrance brothers and one-time mustanger Monty Roberts. These skills have largely come to public notice through Roberts' best-selling *The Man Who Listens to Horses* and Nicholas Evans' 1995 novel, *The Horse Whisperer,* a movie version of which starred Robert Redford.

It can be argued that the increasingly sympathetic portrayal of wild horses in popular culture has raised the level of societal awareness and has influenced public opinion, which, in turn, has shaped the protective legislation and more humane approaches toward mustangs in recent years. It must also be said, however, that few of the depictions in the popular media have addressed the core social, political and scientific issues confronting the continued existence of wild horses in the 21st century West. But they have told us much about ourselves, our inner need for things wild and our hopes for the mustangs of the future. ■

John Eastcott
and Yva Momatiuk

5 SOURCES AND SELECTED READINGS

B. Byron Price

Amaral, A. A. 1980. Will James: T*he Last Cowboy Legend.* Reno, NV: University of Nevada Press.

Ashton, D. 1981. *Rosa Bonheur: a Life and a Legend.* New York: Viking Press.

Audubon, J. J. 1897. A*udubon and His Journals with Zoological and Other Notes by Elliott Coues.* By M.R. Audubon. New York: Charles Scribner's Sons.

Ayres, L. 1987. "William Ranney." In *American Frontier Life: Early Western Painting and Prints,* by R. Tyler et al., with an introduction by P. Hassrick. Fort Worth, TX: Amon Carter Museum; New York, Abbeville Press.

Bell, W. G. 1987. *Will James: The Life and Works of a Lone Cowboy.* Flagstaff, AZ: Northland Press.

Bonheur, R. 1998. *Rosa Bonheur: All Nature's Children.* New York: Dahish Museum.

Bramlett, J. 1987. *Ride for the High Points: The Real Story of Will James.* Missoula, MT: Mountain Press Publishing Company.

Brooks, G. 1990. "Wild Horses" on *No Fences* [sound recording]. Capitol Records.

Buffalo Child Long Lance. 1928. *Long Lance.* New York: Cosmopolitan Book Corporation.

Buscombe, E, ed. 1988. The BFI Companion to the Western. New York: Atheneum.

Cummins, D. D. 1980. *William Robinson Leigh, Western Artist.* Norman: Tulsa: University of Oklahoma Press; Thomas Gilcrease Institute of American History and Art.

Dobie, J. F. 1952. *The Mustangs.* New York: Little Brown.

DuBois, J. 1977. *W. R. Leigh: the Definitive Illustrated Biography.* Kansas City, MO: Lowell Press.

Evans, N. 1995. *The Horse Whisperer.* New York: Delacorte Press.

Frazier, D. 1998. *The Will James Books: A Descriptive Bibliography for Enthusiasts and Collectors.* Long Valley, NJ: Dark Horse Associates.

Garfield, B. 1982. *Western Films: A Complete Guide.* New York: Rawson and Associates.

Harbury, M. 1996. *Last of the Wild Horses.* Toronto: Key Porter Books.

Hardy, P. 1983. *The Western.* New York: William Morrow and Company.

James, W. 1931. *Cow Country.* New York, London: C. Scribner's Sons.

Lawrence, E. A. 1981. The white mustang of the prairies . *Great Plains Quarterly* 1:81-94.

——. 1982. *Rodeo: An Anthropologist Looks at the Wild and the Tame.* Knoxville, TN: University of Tennessee Press.

Leigh, W. R. 1933. *The Western Pony.* New York: Harper & Brothers.

Melville, H. 1930. *Moby Dick.* New York, Modern Library.

Murphey, M. M. 1994. "Running Shadow" on *America's Horses* [sound recording]. Warner Brothers Records.

McLaughlin, C. 1991. Badlands broomtails: the cultural history of wild horses in western North Dakota. *North Dakota History: Journal of the Northern Plains* 53:2-19.

Roberts, M. 1997. *The Man Who Listens to Horses.* New York: Random House.

Rolling Stones. 1995. "Wild Horses" on *Stripped* [sound recording]. Virgin Records America.

Russell, D. 1960. *The Lives and Legends of Buffalo Bill.* Norman, OK: University of Oklahoma Press.

Savage, W. 1979. *The Western Hero: His Image in American History & Culture.* Norman, OK: University of Oklahoma Press.

Shriver, R. 1982. *Rosa Bonheur: With a Checklist of Works in American Collections.* Philadelphia: Art Alliance Press.

Smith, D. B. 1982. *Long Lance: The True Story of an Imposter.* Lincoln, NE: University of Nebraska Press.

Spragg, M, ed. 1997. *Thunder of Mustangs.* San Francisco: Sierra Books.

Troccoli, J. Carpenter. 1990. *Alfred Jacob Miller: Watercolors of the American West:* from the collection of Gilcrease Museum. Tulsa, Oklahoma: Thomas Gilcrease Museum Association.

U2. 1991. "Who's Gonna Ride Your Wild Horses" on *Achtung Baby* [sound recording]. Island Records, Ltd.

The Wild Free-Roaming Horse and Burro Act of December 15, 1971. PL 92-195, 85 Stat 649.

Wilkinson, A. 1999. "The Mustang Song." Lubbock, TX: Caint Quit Music.

About the Author

B. Byron Price has served as executive director of the Buffalo Bill Historical Center in Cody, Wyoming, since 1996. Before coming to Cody, he served in a number of museum and history related positions, including executive director of the National Cowboy Hall of Fame in Oklahoma City, executive director of the Panhandle-Plains Historical Museum in Canyon, Texas, and curator of history at the Panhandle-Plains Historical Museum. A prolific writer, Price is the author and co-author of numerous books, monographs and articles on the history and culture of the American West. Price is a frequent speaker at professional meetings and symposia.